An open-pit copper mine in Utah

Copper

Peter Murray

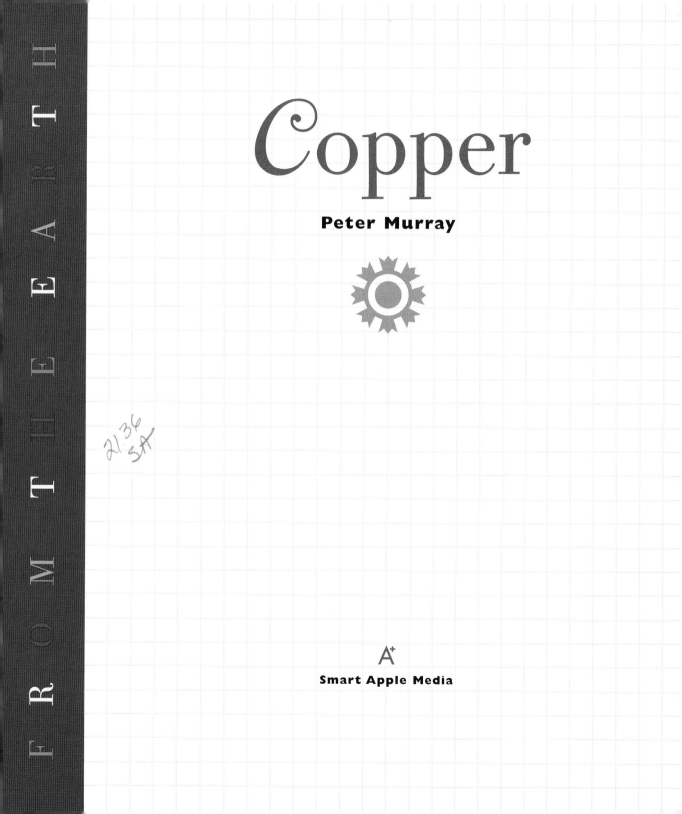

2136
SA

A⁺
Smart Apple Media

COPYRIGHT

☼ Published by Smart Apple Media

1980 Lookout Drive, North Mankato, MN 56003

Designed by Rita Marshall

Copyright © 2002 Smart Apple Media. International copyright reserved in all countries. No part of this book may be reproduced in any form without written permission from the publisher.

Printed in the United States of America

☼ Photographs by Frank Balthis, JLM Visuals (Richard Jacobs), Jeff Myers, Tom Myers, Richard Nowitz, Rainbow (Dan McCoy)

☼ Library of Congress Cataloging-in-Publication Data

Murray, Peter. Copper / by Peter Murray. p. cm. — (From the earth)

Includes bibliographical references and index.

☼ ISBN 1-58340-112-1

1. Copper—Juvenile literature. [1. Copper.] I. Title. II. Series.

TN780 .M87 2001   669.3—dc21   00-069820

☼ First Edition 9 8 7 6 5 4 3 2 1

Copper

CONTENTS

The Copper Age   6

Copper Alloys   12

Mining Copper   14

Copper Today   18

Hands On: Make a "Copper Nail"   22

Additional Information   24

The Copper Age

In 1991, high in the Alps, a dead man was found frozen in the ice. His clothes were made of leather and grass. His body was marked with strange tattoos. He became known as "the Iceman." ❋ The Iceman had a backpack made of birch bark, a fur quiver full of handmade arrows, and a flint dagger. He also had a strange wooden ax with a small copper blade. This ax told scientists that the Iceman probably died about 5,000 years ago, during the Copper Age. ❋

Copper ore

For many thousands of years, copper was the only metal widely used by humans. Long before the mining of gold, silver, or iron, copper was made into jewelry and tools. ☀

Pure copper is a soft metal. It can be hammered and bent into almost any shape. The ancient Egyptians polished sheets of copper to make mirrors. Inside the pyramids, archaeologists have found copper drinking cups, tweezers, dishes, razors, and dozens of other copper items.

A copper pendant more than 10,000 years old was discovered in Iraq.

Copper pipes and fittings

The inside of a copper mine

Copper Alloys

Around 2500 B.C., metalworkers learned that when they melted copper with small amounts of **tin**, they got a harder, gold-colored alloy called bronze. For 2,000 years, bronze was the most common metal used for tools, hardware, and **cast** metal sculptures. ✺ Mixing **zinc** with copper produces brass, the hard yellow metal used in trombones and other musical instruments. Copper is used in many other alloys as well, and

America's famous Liberty Bell is cast bronze.

The Liberty Bell in Philadelphia, Pennsylvania

each alloy has its own special uses. Our coins are made

of copper alloys—even the coins that look like silver!

Mining Copper

In nature, most metals are mixed with rock. This mixture

is called ore. But copper is one of the few metals that can also

be found in its pure form on the surface of the earth. Large

nuggets of **native copper** were once common. The biggest

copper nugget ever found weighed more than one million

pounds (450,000 kg)! ☀ After thousands of years of mining,

most of the native copper nuggets are gone. But copper ore is

plentiful. Most copper ore is less than one percent copper.

Millions of tons of copper ore are dug up from huge open-pit

mines each year. The ore is ground into fine sand, then tiny

Earth that was mined for copper

bits of copper are separated from the rock. The copper is

melted and purified in a process called **smelting**. It is sold in

bars of pure copper called ingots. Because it is a very good

electrical conductor, most of the copper **Before 1982, pennies were 95 percent**

mined today is made into electrical wire. **were 95**

percent

It is also molded into pipe, sheeting, and **copper and 5 percent**

5 percent

hundreds of other items. Every year, about **zinc.**

two million tons (1.8 t) of copper are mined in North America.

Copper wires inside a computer chip

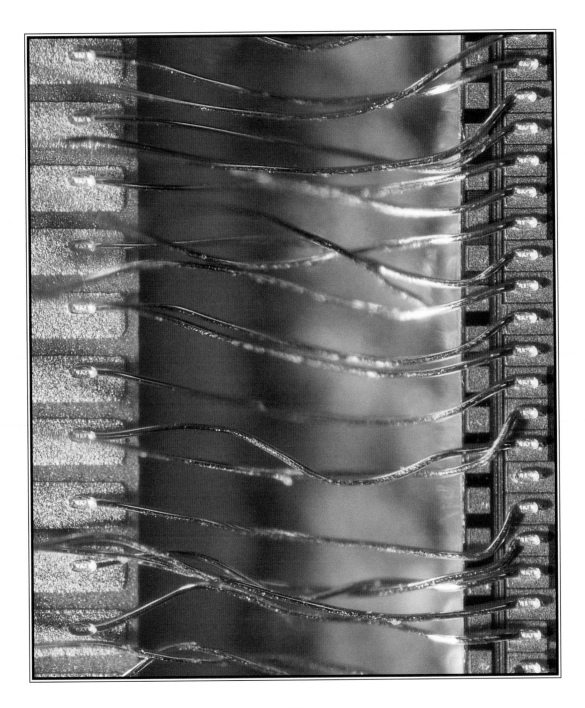

Copper Today

Ten thousand years after the first metalworker pounded a copper nugget into something useful, copper is still an important part of our everyday lives. **Over time, exposure to sunlight and weather gives copper a green color.**

Because it is very durable, sheets of copper are sometimes attached to the roofs of buildings. Every time someone flips a light switch, copper wire carries the electricity to the light bulb. Copper pipes carry water to faucets. Colorful copper compounds are

A church dome covered by copper

18

used to dye clothing and fertilize crops. Copper can be found in computer circuits, automobile radiators, electric motors, and the pots and pans we use for cooking. Artists shape copper into sculptures and monuments such as the Statue of Liberty. ☀

Copper can be recycled as well. Each year, we use just as much recycled copper as copper from newly mined ore. Without copper, our lives would be very different. It is one of the world's most useful and important metals.

Copper ore nuggets

Make a "Copper Nail"

What You Need

A glass

1/2 cup (120 ml) vinegar

1 teaspoon salt

10 old (not shiny) pennies

An iron or steel nail

Steel wool

What You Do

1. In the glass, mix the vinegar with the salt. Put the pennies into the glass and stir them around for about a minute.

2. Clean the nail with the steel wool. The nail must be very clean.

3. Put the nail in the glass with the vinegar, salt, and pennies and wait one hour.

What You See

The nail has turned the color of copper. The acid in the vinegar dissolved a thin layer of copper from the pennies. The dissolved copper then stuck to the iron in the nail. You now have a copper-plated nail!

Coins made of copper alloys

INFORMATION

Index

alloys 12, 14

brass 12

bronze 12

Copper Age 6

Iceman 6

Liberty Bell 12

mining 14–16

ore 14–16

uses 8, 12, 16, 18, 20

Words to Know

alloy (AL-oy)—a mixture of two metals

cast (KAST)—to pour molten metal into a mold

native copper (NAY-tiv KAH-pur)—naturally occurring copper nuggets

smelting (SMEL-ting)—melting ores to extract metal

tin (TIN)—a soft, silver-colored metal

zinc (ZINK)—a brittle, bluish-white metal

Read More

Bates, Robert. *Mineral Resources A-Z.* Hillside, N.J.: Enslow, 1991.

Knapp, Brian. *Copper, Silver and Gold.* Danbury, Conn.: Grolier Educational, 1996.

Lessen, Don. *The Iceman.* New York: Crown, 1997.

Symes, R. F. *Rocks and Minerals.* New York: Knopf, 1988.

Internet Sites

The Copper Page

http://www.copper.org

Sixty Centuries of Copper

http://www.60centuries.copper.org/

All About Pennies

http://www.pennypage.com/about_pennies.htm